纽约建筑室内设计

Interior design of New York architecture

深圳视界文化传播有限公司 编

中国林业出版社

图书在版编目（CIP）数据

纽约建筑室内设计 / 深圳视界文化传播有限公司 编.
-- 北京：中国林业出版社，2018.7
ISBN 978-7-5038-9321-6

Ⅰ.①纽… Ⅱ.①深… Ⅲ.①室内装饰设计－纽约
Ⅳ.①TU238.2

中国版本图书馆CIP数据核字(2017)第254154号

编委会成员名单

策划制作：深圳视界文化传播有限公司（www.dvip-sz.com）
总 策 划：万绍东
编　　辑：丁　涵
装帧设计：潘如清
联系电话：0755-8283 4960

中国林业出版社 · 建筑分社
策　　划：纪　亮
责任编辑：纪　亮　王思源

出版：中国林业出版社
（100009 北京西城区德内大街刘海胡同7号）
网站：http://lycb.forestry.gov.cn
电话：（010）8314 3518
发行：中国林业出版社
印刷：北京利丰雅高长城印刷有限公司
版次：2018年7月第1版
印次：2018年7月第1次
开本：1/16
印张：20
字数：300千字
定价：428.00元

尽管室内设计努力朝创意方向发展，但仍然要符合惯例。这些惯例会随时间推移不断调整，然而并不是所有人都赞同其特质。新纪元充满创意与机遇，在任何给定时间都可以对这些类型加以阐述，不同的风格、基调、色彩运用得恰到好处，人们希望看到的一切都承载着良好的品位。将这些不同风格取其精华、去其糟粕，以创造性的眼光，最大程度加以利用，需要了解设计历史。

当代室内设计谈及当下，即此时此刻。新型材料、最新科技、创造能力、意想不到的色彩以及非常规运用等，并非所有东西都融洽地应用于现代家庭之中，但这是立足于当下的理念。客户和设计师要运用它们，需要活力、勇气与热情。

如今的现代室内设计被经受住时间考验的多重元素影响。这些创意与理念源自20世纪中叶，与该时代艺术文化变革同步。更进一步思考，它们现在已成为了设计传统的一部分。这些现代室内设计在生机勃勃与轻描淡写、色彩斑斓与单色渲染、奢华精致与质朴无华之间不断变换。房间不一定都有特殊用途。大部分家居线条利落，有硬角也有软包，非常整洁，没有任何多余饰物。表面材料无论亮面还是哑光，都很平滑流畅。织物经过创新改造，仍然舒适实用。现代室内设计依赖于细节，当未装饰部分很多时，则需要更巧妙地去完成。

18及19世纪室内设计大多都与传统室内设计，尤其是家居方面有关。装饰元素的安置和点缀非常重要，但这并不意味着每个表面、每扇窗都要经过抛光、软装及褶皱状帘布覆盖。每个房间都有具体的用处，比如门厅/走廊、客厅、餐厅、图书室、凉廊、书房等。根据墙壁、门以及嵌入的部分，合理安排建筑细节。色彩、质地及样式的使用或许很复杂，却也可以变得简单，缩减至单一色彩或样式。现当代艺术不是某一时代的特定产物，所以将被运用于传统房间之内。

一个过渡的室内空间可以既现代又传统，从而使它变得舒适、温馨且易亲近。建筑元素或许会为室内家居表明方向，但不能仅局限于单一规则。设计师对早期阶段的设计理念加以重新诠释和总结，他们引用常见纹理图案的织物以新的尺寸及颜色进行渲染。

如果有效地运用折衷方式，这些元素就会像精致食谱中各种各样的配料一样成功地汇集在一起。对比色创建了对话，但是最好维持在同一色系，使用一个对比色突出该色系。应该用一个显著特色加上备选形状、表面材料以及图案来显示彼此之间的关联性，突出明显的对比。

如今的室内设计不局限于任何一种界定的类型，因此书中所描述的各个空间也同样不受约束。特殊的家具、材料甚至是不能定义类型或风格的家具，工具和美学互相促进，其整体运用成为决定性的因素。拥有关注的愿景非常重要，而最重要的是拥有想象力。

CONTENTS 目录

006
Stephen Mitchell
Designlush
SUMMER VACATION HOME
夏日度假别居

022
Carlton Holmes, Kami Cheney
Holmes, King, Kalquist & Associates, Architects, LLP
AN EXQUISITE LODGE NEAR THE BIG MOOSE LAKE
大穆斯湖畔的精美木屋

046
John Barman
John Barman Inc.
COLORFUL WORLD
炫彩世界

064
Jack Ovadia
Ovadia Design Group
THE BEAUTY OF MODERN LUXURY
现代奢华之美

082
Stephen Mitchell
Designlush
THE BEAUTY OF NATURE
自然之美

102
Jamie Drake
Drake Design Associates
METROPOLITAN CHARM
都市风情

112
Ghislaine Viñas, Vané Broussard, Hayley Singleton
Ghislaine Viñas Interior Design
SKYHOUSE
天际

122
Jamie Drake
Drake Design Associates
THE STARLIGHT
星光

134
Keith Baltimore
Baltimore Design Group
COLORFUL FAMILY RESIDENCE
缤纷家居

154
Breanna Carlson & Brynn MacDonald
bSTUDIO Architectural Design, LLC
THE ELEGANT RUSTIC HOME
乡村雅居

168
Suzanne Caldwell, Maria Greenlaw
Design House
DELICATE COTTAGE
精致小屋

178
Andrew Franz and Jaime Donate
Andrew Franz Architect PLLC
HOME AND NATURE
自然与家

186
Sara Story
Sara Story Design

WALKING ALONG THE BEACH
漫步沙滩

198
Fawn Galli
Fawn Galli Interiors

THE DANCING FAIRY
舞动的精灵

206
Fawn Galli
Fawn Galli Interiors

BACK TO NATURE
回归自然

228
John Willey
Willey Design LLC

CALM AND ELEGANT
静谧与优雅

236
Breanna Carlson, Brynn MacDonald and Tilton Fenwick
bSTUDIO Architectural Design, LLC

A LITERARY JOURNAL
艺术之旅

248
Suzanne Caldwell, Maria Greenlaw
Design House

RETRO AND MINIMALIST
复古与简约

256
John Willey
Willey Design LLC

BRANDISH YOUTH
舞动的青春

268
Jeff Pfeifle

GORGEOUS SPACE
华丽空间

278
Joe Nahem and David Gorman
Robert Couturier Inc.

A CHARMING TEMPORARY HOME
魅力别居

286
Cathy Triant Buxton
Cathy Triant Buxton T.D.Triant Inc.

THE HERITAGE OF LOVE
爱的传承

302
Joe Nahem and David Gorman
Fox-Nahem Associates

A WARM AND COMFORTABLE RESIDENCE
温情家居

312
Geoffrey Bradfield
Geoffrey Bradfield, Inc.

THE GLAMOUR OF ART
艺术之魅

NEW YORK LIVING · AMAGANSETT, NY

SUMMER VACATION HOME

夏日度假别居

设计公司
DESIGN COMPANY
Designlush

设计师
DESIGNER
Stephen Mitchell

摄影师
PHOTOGRAPHER
Arthur Keller

This home is designed as a summer retreat with the intention of escaping the busy "go go go" city mentality. It was requested that the space be minimal, contemporary, comfortable and set up in a way to entertain on a whim. They also wanted to be able to enjoy the outdoor space during all times in all weather conditions—therefore multiple outdoor lounge areas were created (Pool-side, outdoor covered terrace, outdoor kitchen and private outdoor lounge areas off each main room).

本案设计为一处避暑佳所，意图逃离城市忙碌的生活节奏。屋主要求空间以不经意的方式呈现简约、现代与舒适。他们还希望一年四季都能欣赏室外景色，因此设立了多重室外休闲区（泳池旁的遮阳台、户外厨房以及每个主要房间外的私人休闲区）。

As this is a summer vacation home the designer wanted to create individual indoor + outdoor settings each designed for a specific purpose that all achieved the main goal of creating an inviting, relaxing and comfortable moment. Whether you want to prepare a fantastic meal, have a glass of wine in front of the fireplace or catch up on your latest novel, you can do it all in a comfortable and inviting indoor or outdoor environment. In addition the designer wanted areas that were built to be personable yet also set up to entertain and hanging-out.

设计师想把这个夏日度假小屋设计成为室内外各自独立的布局，设想虽不相同，但同样意欲打造好客、休闲、舒适的环境。不论你是想准备一顿丰盛的晚宴，在壁炉前品一杯红酒，亦或是补看最近更新的小说，在这个舒适好客的室内外环境中你可以做你想做的一切。此外，设计师希望能设计有吸引力的地方，既可以休闲娱乐，又可以四处闲逛。

The designer set out to create an architecturally modern + relaxing vacation home that easily transitions from usable indoor to outdoor space. The home is 780 square meters with six bedrooms and six baths. The house offers two master suites and the kitchen, fireplace, private lounge areas and two private outdoor showers.

我们意图打造一个在建筑角度上现代化、休闲化的度假小屋，可以轻易从室内空间过渡到室外。房屋面积780平方米，有6间卧室及6间浴室。屋内有两间主卧套房、厨房、壁炉、私人休息区，还有两间私人户外淋浴房。

NEW YORK LIVING · BIG MOOSE LAKE, NY

AN EXQUISITE LODGE NEAR THE BIG MOOSE LAKE

大穆斯湖畔的精美木屋

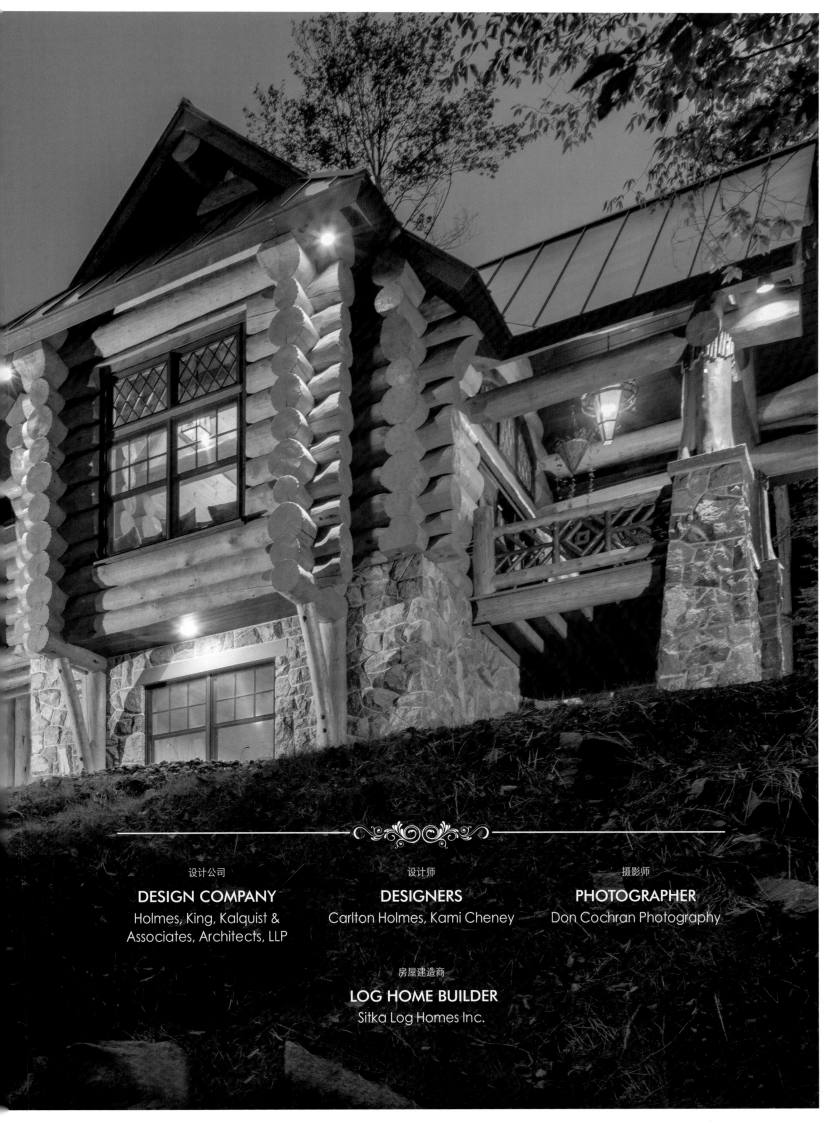

设计公司
DESIGN COMPANY
Holmes, King, Kalquist & Associates, Architects, LLP

设计师
DESIGNERS
Carlton Holmes, Kami Cheney

摄影师
PHOTOGRAPHER
Don Cochran Photography

房屋建造商
LOG HOME BUILDER
Sitka Log Homes Inc.

This Swedish Cope style scribed log lodge is perched on a hillside overlooking Big Moose Lake.

The landscape provides some privacy from the lake using the existing trees, yet still offers an appesing view out to the water and frames the beautiful lake views. The land was left in its natural state with a mix of softwood and hardwood trees, as well as woodsy shrubs and ferns. There are no lawn areas.

这个瑞典风格的木屋位于可远眺大穆斯湖的一座山上。它巧妙地利用湖边树木成荫,但同时也呈现出美丽的湖景。这片土地保留着它原始的状态,培育着软木和硬木树种以及灌木和蕨类植物,这里没有草坪。

- The hardwoods from the site were sent to a mill for preparation to be used for all of the flooring and cabinetry.

  此处的硬木树经过工厂加工后用于制作地板和橱柜。

- The bedrock encountered was removed by blasting and then re-used for the piers, foundation cladding, fireplaces and chimneys.

  建造过程中所遇到的基岩经爆破后被移除,它们被重新用于制作支墩、地板覆盖层、壁炉以及烟囱。

Sitka Log Homes utilized standing dead trees, with relatively low moisture content for the log work. This allowed the architects to design for minimal settlement. The log building was handcrafted and assembled at the Sitka Log Homes construction yard. When completed, each log was number tagged, disassembled and loaded onto semi-trailers for transport to the home site in New York State.

建造商Sitka Log Homes公司就地取材，利用水分含量较低的枯木进行建造，这也大大减少了这项工程的预算。这些建筑材料在Sitka Log Homes公司的建造工厂进行手工制作和组装，组装完成后每根原木都会被贴上编号，装上半挂车后运往目的地纽约州。

- The log assembly was created using 16" diameter logs of Engelmann spruce not only on the exterior, but all interior walls as well.

  直径为16英寸（约40厘米）的英格曼云杉运用到了木屋的室外和内墙装饰中。

- The flair butts used for the decorative post and beams were harvested using Western red cedar, with diameters ranging up to 30".

  用于装饰的柱子和横梁都取材于直径为30英寸（约76厘米）的西红杉。

- The kitchen and bathroom cabinetry was created by local craftsmen using hardwood stock made from the harvested timber and was crafted in a traditional bark and stick motif.

  厨房和浴室内的橱柜是请当地工匠用丰收的木材用传统方式进行加工的。

- The lighting also was locally sourced with a manufacturer located in the Utica area having an outlet in Old Forge.

    照明设备也是由尤蒂卡地区的制造商在当地Old Forge的零售商店所提供的。

- The standing seam roofing was crafted on-site using coil stock.

    立接缝屋面是用卷料进行现场加工的。

- The building is heated on both the lower level and main level using in-floor radiant tubing.

    地板内的辐射管的运用使木屋两层都可以取暖。

- From the garage through to the master suite, the entire first level is accessible, including roll-in shower and doorways.

    从车库到主卧套房，包括浴室和门廊，整个一楼都是相通的。

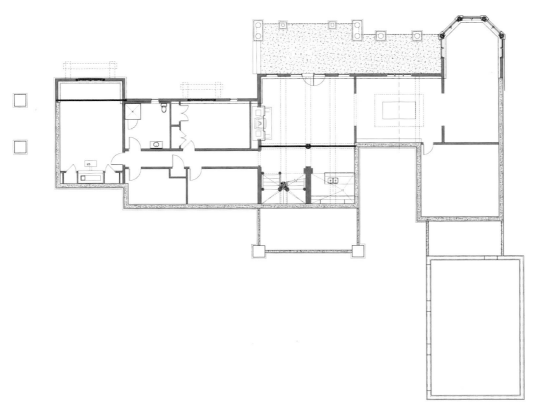

The main level has a two-story great room flanked by the master suite with laundry included to the left and kitchen, dining and family mud room entry from the garage to the right. The winding stair takes you to the loft overlooking the great room and below to the lower "walk-out" family room that leads to the beautiful lake.

木屋两层都有客厅，客厅侧面是主卧套房，套房左侧有洗衣机，车库右侧依次是厨房、餐厅和房间入口。旋梯可以将你带到木屋的顶层，其间你可以俯瞰通往美丽湖畔的客厅和家庭房。

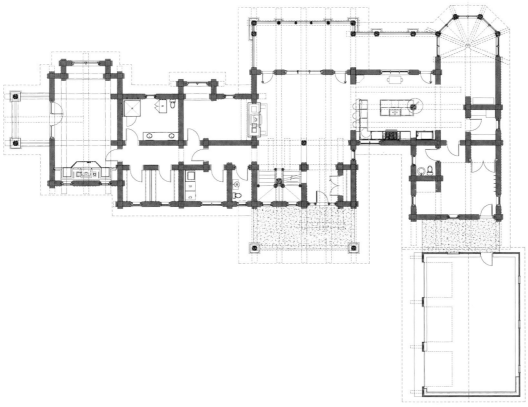

NEW YORK LIVING · NEW YORK CITY, NY

COLORFUL WORLD
炫彩世界

设计公司	设计师	摄影师
DESIGN COMPANY	**DESIGNER**	**PHOTOGRAPHER**
John Barman Inc.	John Barman	Anastasios Mentis

This is a duplex penthouse apartment in New York City which is renovated completely by the designer. It is a pre-war building and the designer restored the moldings and installed a new staircase. The lady of the house likes colors, thus the designer chose a vibrant palate reminiscent of Venice.

The living room was designed to showcase the client's collection of modern art and the colors of the room were inspired by the paintings. Arched French doors open on to the large terrace.

本案为一栋位于纽约市的复式公寓。它是一栋战前建筑，设计师重塑了它的外观并安装了新楼梯进行改造。这里的女主人喜欢色彩，所以设计师选择了能让人联想起威尼斯的活力配色。

客厅的设计展示了客户的现代艺术品收藏，室内颜色受油画的灵感启发。拱形落地门开向大型平台。

The solarium has a 20 feet sofa which faces the tall wall of bookcases and is great for watching TV.

日光房中有一张20英尺长的沙发,面朝高大的书橱,坐在沙发上看电视想必非常棒。

The study is warm and inviting and can also serve as a guest room.

The pink entry hall has a custom floor and a painting by Kelly Stuart Graham on the stairs.

书房温馨好客，还可以作客房用。

粉色入口门厅采用定制地板，楼梯墙壁上挂着凯利·斯图亚特·格雷厄姆的画作。

The master bedroom has a mirrored fireplace and French doors that open on to a private terrace. The silver wallpaper with the blue floral design was the inspiration for the room.

主卧室内有一座镜子般的壁炉，落地门开向私人平台。银色壁纸及蓝色花纹设计是这个房间的灵感所在。

The rich and vibrant colors were inspired by the colors in the client's contemporary art collection. The clients wanted European feeling of warmth and elegance and luxury in her New York duplex penthouse. The 20 feet long blue sofa in the book lined solarium is covered in an indoor-outdoor fabric for practicality. Contemporary upholstery has been mixed with mid-century furniture and antiques to complete the look for the apartment.

This apartment is elegant and comfortable. Cozy in the winter and wonder in the warm weather with the terraces and solarium with plenty of seating for entertaining.

丰富而有活力的色彩是受客户当代艺术收藏品的启发。客户希望在她的纽约复式公寓中拥有欧式的温馨、优雅与奢华。日光房中20英尺长的蓝色沙发上覆盖着一条室内外皆宜的织物，增加了实用性。公寓的设计融合了当代风格软装饰品与中世纪的家具、古董。

本公寓典雅舒适，在冬日里温馨惬意，天气暖和时奇妙非凡，露台及日光房里有许多休闲落座区。

NEW YORK LIVING · BROOKLYN, NY

THE BEAUTY OF MODERN LUXURY

现代奢华之美

设计公司
DESIGN COMPANY
Ovadia Design Group

设计师
DESIGNER
Jack Ovadia

摄影师
PHOTOGRAPHER
Francis Dzikowski

The design concept here is to make the family and homeowner feel as if they were in a luxury hotel when they will come home every day. The style dated back to a modern version of the Great Gatsby era where luxury, sophistication and glamour were at the front most and was expressed through every single material and furnishing of the home. The house with feel constant moving from one room to another, with connecting materials and color pallets, but feel overall airy and light but also functional for a day-to-day growing family. The circular stair is a main focal point, which connects you from floor to floor, with each floor serving a specific feature. The basement is for kids playroom and fitness center where things can get messy, the first floor is for entertainment with the public rooms, the second floor is for sleeping and privacy where the bedrooms are situated and the third floor is for the guest who visits.

本案设计理念是令屋主及家人每天回到家里都能感觉仿佛置身于豪华酒店。其风格是伟大的盖茨比时代的现代版本，豪华、精细与魅力展现在每一种材质及家居之中。在每个房间都可以有"家"的感觉，材质及配色都互相连接，整体通风及采光良好，对一个日常生活的家庭来说功能齐备。圆形楼梯是一个主要焦点，连接各层楼，每一层都有专属特征。地下室用作小孩游戏室及健身房，这里的东西可以任意摆放，一层是带有公共房间的娱乐室，二层是具有私密性的卧室空间，三层是为访客准备的。

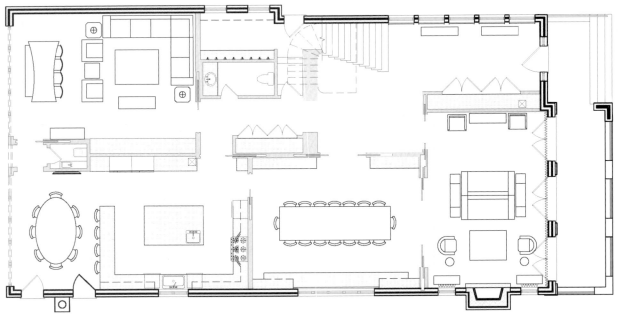

The colors and materials used for the home are of elegance and richness. The common area is a custom wall panel, in a off-white lacquer trimmed in bronze and the inner layer was quartered oak, bleached and painted in a gloss finish and accented with Eramosa marble slab flooring. This feature runs throughout the house connecting each floor and is the guide throughout the public spaces. When going from room to room, each has its materials of lush velvets, crystal custom chandeliers, wool and silk carpets, accented emerald green and walnuts woods and fine fabrics. The rooms are dressed with fine artwork and accessories for completion of the space. The kitchen is decorated with grained gloss walnut with porcelain floors for durability. Each room has a custom-made millwork unit which accommodates the functions of that room, but also acts as focal point artwork piece. Like the one in the family room, it has a angular shelf design for books and an octagon trim design to house the TV. In the bedrooms, each room is specifically designed to accommodate the children, from the growing boy, to the family princess and the new born. Each is outfitted with their own tailored custom bathroom as well.

家中选用的色彩及材料雅致而丰富。公用区定制墙板采用米白色漆面镶嵌黄铜，内层材料则是四开橡木经漂染抛光而成，与Eramosa大理石地板一并得以强调。该特征遍及整个室内空间，不仅连接每层楼，还可作为公共区域的指引。从这间房走到另一间房，你会看到奢华的丝绒、定制水晶吊灯、羊毛及丝绸地毯、鲜绿色胡桃木地板及精致布料。房间以精致艺术品及配饰作为装饰，填充整个空间。厨房采用木纹光面胡桃木加瓷地板，经久耐用。每个房间都有定制木工，不仅与室内功能相适应，而且起到了画龙点睛的作用。家庭活动室内有一个有角书架，八边形装饰的设计方便放入电视。每间卧室都为孩子们特殊打造，有男孩房、女孩房和婴儿房，还配备了量身定制的浴室。

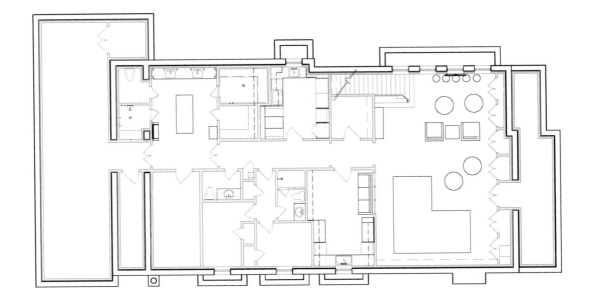

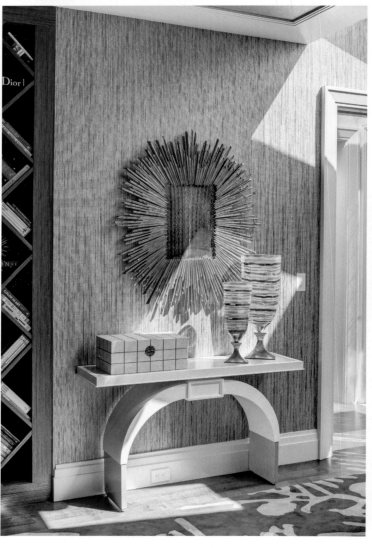

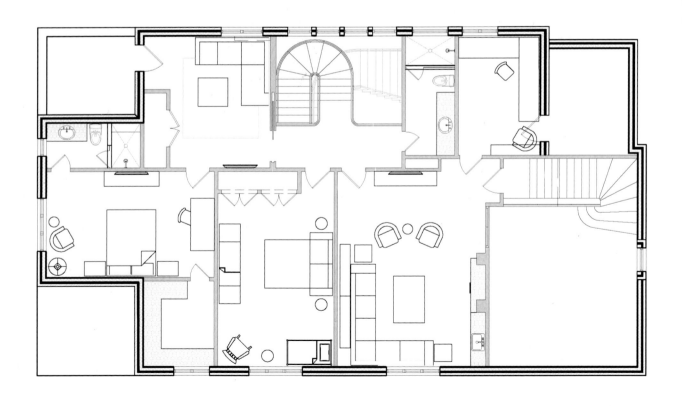

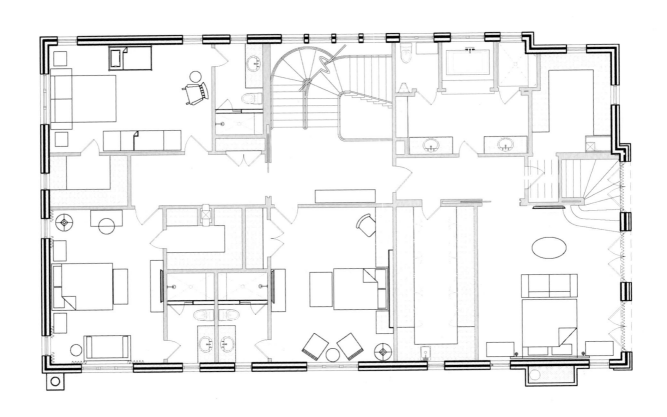

081

THE BEAUTY OF NATURE
自然之美

设计公司	设计师	摄影师
DESIGN COMPANY	**DESIGNER**	**PHOTOGRAPHERS**
Designiush	Stephen Mitchell	Bill Timmerman, Zack Hussain

087

The Mojave landscape maintains an inherent beauty of textures, stratifications and materials as well as protected oasis of color brought to life under the play of shadow patterns of a harsh sun. These environmental realities can be used as inspiration for design to create a sense of place and character regionalism. In this project, the materials develop a layering of mass as you move from the basement to the private realm. The interiors were chosen to compliment the architectural aesthetic of the home and it's organic open feel. The space is designed around the families way of life.

莫哈维沙漠的自然景观有与生俱来的美丽的纹理、层次和材质，也有能为酷日提供一丝阴凉的绿洲。这些自然环境中的现实条件为创造一个有特色的住宅提供了灵感。这个项目中，材料的分层使用从地下室到私人区随处可见。内饰的选择衬托了这个房子的建筑美感，整体上有一种自然开放的感觉。整个空间的设计围绕着这个家庭的生活方式展开。

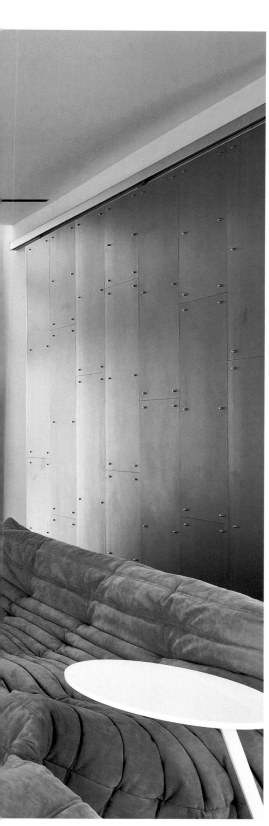

097

NEW YORK LIVING · TRIBECA, NY

METROPOLITAN CHARM
都市风情

设计公司
DESIGN COMPANY
Drake Design Associates

设计师
DESIGNER
Jamie Drake

摄影师
PHOTOGRAPHER
Marco Ricca

A sprawling loft in a converted industrial building provided the canvas for this Jamie Drake design. The duplex apartment has a large outdoor space, with a dynamic flow of the volume; its skylights create an energetic movement. The vast living space on the lower floor was modulated with level changes, creating intimate zones in what could have been an overly cavernous space. Vibrant, intense colors further warm the rooms. Rich architectural materials of travertine, wood, gilt glass, steel, plaster, lacquer and quartz are married with leather, silk and wool textiles in bold fashion.

这间由工业建筑改造而成的豪华复式公寓是杰米·德雷克设计公司的作品。它拥有宽敞的室外空间，其天窗打造出一种活力四射的动感。设计师对下层广阔的客厅空间进行楼层改动，那个洞穴般幽暗的空间被改造成私密区域。生机勃勃、热情洋溢的色彩进一步使这些房间变得温馨。石灰石、木材、镀金玻璃、钢材、塑料、清漆及石英等品种丰富的建筑材料与皮革、丝绸、羊毛纺织品大胆结合，形成设计新风尚。

109

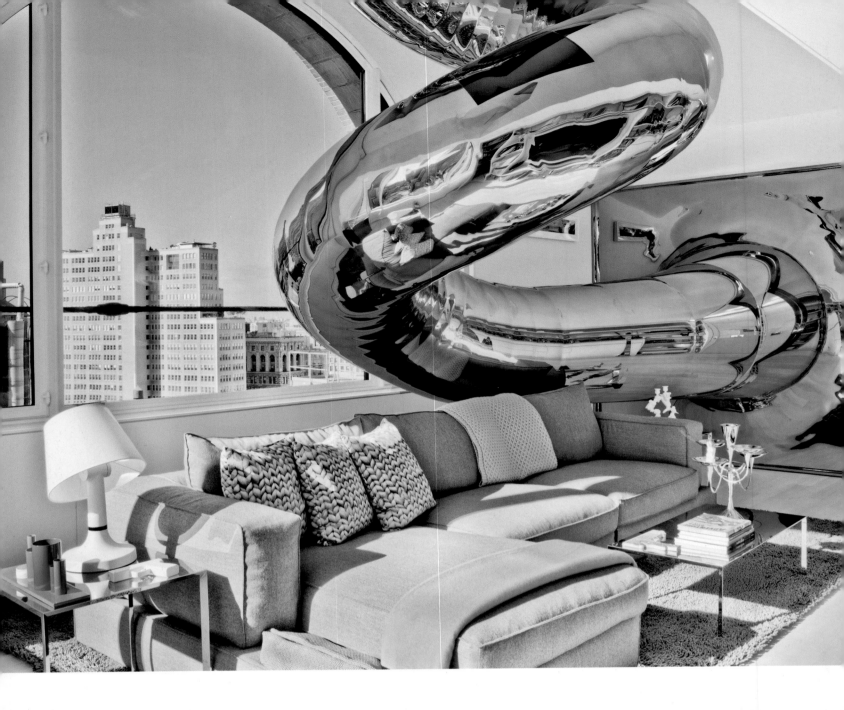

NEW YORK LIVING · NEW YORK CITY, NY

SKYHOUSE
天际

室内设计公司
INTERIOR DESIGN COMPANY
Ghislaine Viñas Interior Design

室内设计团队
INTERIOR DESIGN TEAM
Ghislaine Viñas, Vané Broussard, Hayley Singleton

摄影师
PHOTOGRAPHER
Eric Laignel

Ghislaine Viñas Interior Design was so pleased to be a part of this incredible project for their previous clients. The four story penthouse of this 1895 skyscraper (one of the earliest remaining structures in New York City) was never used as a residence, only serving as the crowning jewel to the building. When purchased by the owners, it was a completely raw space, having only its original steel structure and strangely configured partial floors. David Hotson was brought in to take this massive renovation on as architect, and transformed it into an adult playground complete with a 80' stainless steel slide, intersecting vistas from glass walkways, and a climbing "wall" off one of the original columns.

吉莱纳·维纳设计公司很高兴能参与以前的客户带来的这个不可思议的项目。这套四层公寓位于1895年建成的摩天大楼（纽约市最早的遗留建筑之一）中，先前从未用于居住，一直被视为珍宝。业主购买时，它完完全全是一个毛坯房，只有原始的钢筋结构以及一些奇怪的楼层配置。戴维·霍特森受雇将其重建，并将其改造成为一个成年人游乐场，这里有80米长的不锈钢滑梯，玻璃通道外的远景，还有一个由原始柱子改造而成的可以攀登的"墙壁"。

Viñas's design team brought a warmth to the otherwise austere space, injecting color, oversized prints, pop culture references and more color! The living room is the only area treated completely in whites and creams, allowing the view from the 21st floor to take precedence. The dining room features a custom two-piece corian dining table by UM Project. The breakfast nook next to the kitchen is furnished in bright greens, and features a design by Viñas using the owners collection of vintage and new plates from their previous home, just displayed in a completely new way. Above the custom colored Divis table is the chandelier by London designer Tim Fishlock. A tucked away ladder takes you up to the mezzanine which houses the office and overlooks the living room and breakfast nook. 2 fluorescent orange chairs are perched there for reading and gazing at the view.

维纳设计团队为这个简朴的空间带来了温暖、色彩、超大印花与流行文化。客厅是唯一一个完全为白色和奶油色的空间，从这里可以优先看到21楼的景色。私人定制的可丽耐大理石餐桌餐椅是餐厅的特色。厨房旁的早餐桌是翠绿色的，墙面上以一种崭新的方式装饰着屋主以前的古董收藏及新盘子。定制的彩色戴维斯桌上方悬挂着英国设计师蒂姆·费什洛克打造的水晶灯。隐蔽的梯子带你进入夹楼，夹楼是办公区，可以俯瞰客厅和早餐桌。这里有两把橙色荧光椅，以便阅读和观景使用。

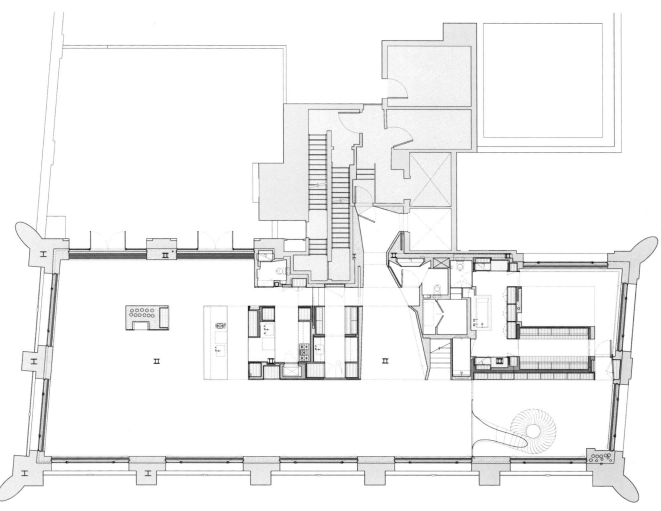

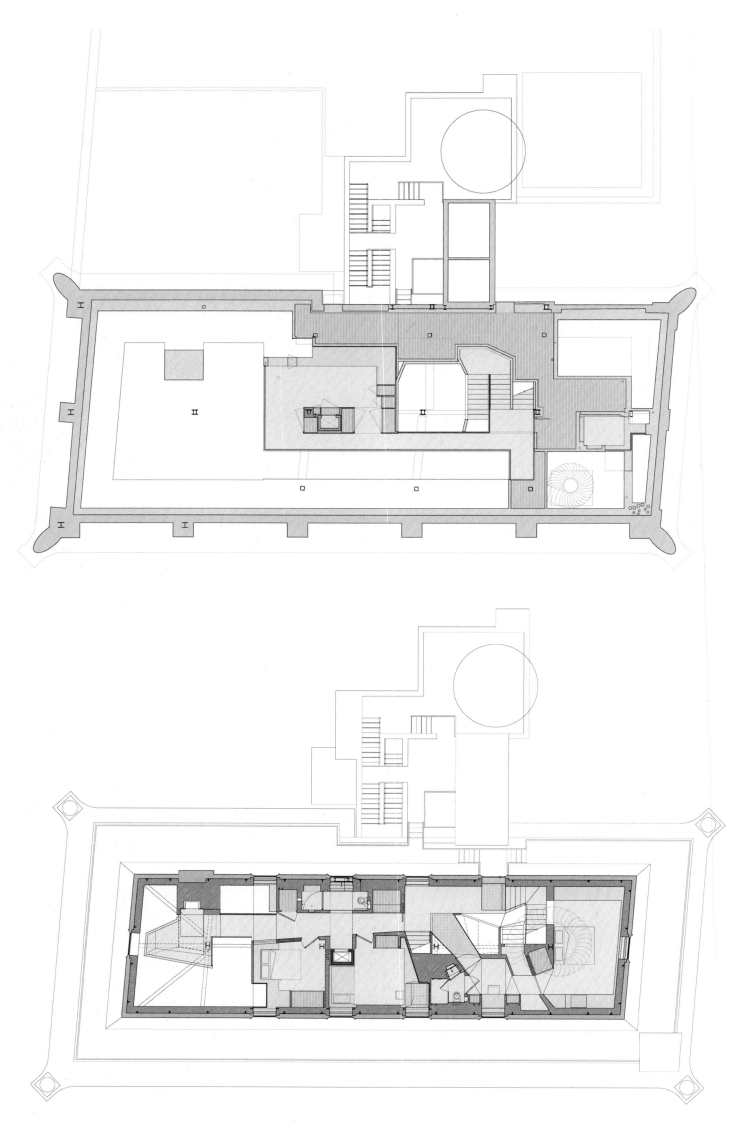

Continuing upstairs are the guest bedrooms, which are accessible by glass walkways, where you can see both above to the upper level and below to the entry. The master guest room features an incredibly whimsical mural by Mexican designers, while the rest of the room is relatively white. From this room you can also access the bottom half of the slide to the 21st floor. Across the walkway is a bathroom with 2 extraordinary features. If you open up the medicine cabinet, you can see straight through to the Brooklyn Bridge, and through a tiny peephole in the shower, you get a clear shot of the Chrysler Building. The mint green guest bedroom has a glass topped desk that brings light down into the stairwell below, and has a lovely view of the Woolworth Building across the way. At the end of the walkway is another glass bridge that leads to the "nest". The floor here is glass so you can see below to the living room.

继续上楼则是客房，由玻璃通道进入，既可以看到楼上，也可以看到楼下的入口。主客房的特色是那幅极其古怪的来自墨西哥设计师的壁画，其他部分都是白色调。这个房间可以通过滑梯到21楼的底部。穿过通道是一间浴室，有两大特色。打开医药柜，能直接看到布鲁克林大桥；通过小窥视孔，可以看到克莱斯勒大厦。薄荷绿的客房有一张能将光引入楼梯间的玻璃桌，这里也能观赏到对面伍尔沃斯大楼的美景。通道尽头是另一座通往"鸟巢"的玻璃桥。地面由玻璃制成，所以可以看到下面的客厅。

The final floor up features the attic space, with cozy nooks for hanging out, porthole window views on either side of the long expanse, and precarious looking glass walls that overlook the living room on one side and the master guestroom on the other. This is also where the top portion of the slide is accessible, which snakes over the master guestroom bed and down through the lower floor library, depositing you just outside the dining room through a wall of stainless steel.

顶层的阁楼有能闲逛的温馨角落，从舷窗窗口可以看到外面的美景，透过不稳定的窥镜墙可以俯瞰客厅与另一边的主客房。这里是靠近滑梯顶端的地方，滑梯沿主客卧的床蜿蜒而下，直至下层的书房，通过一面不锈钢墙壁，直达餐厅外面。

Back on the first floor, to the left of the entry is the master bedroom and library. The library, where the slide hovers precariously over the seating, is a snuggly area featuring a sectional in cushy grey felt and a nubbly rug, also in grey. Accents in turquoise blue are the armchair and throw on the sofa. The master bedroom is somewhat an ode to the city, featuring the City Park wallpaper.

回到一楼，入口左侧是主卧室和书房。书房中，滑梯沿座位盘旋，摇摇欲坠，但这里看上去很温暖，组合家具为灰色系，块状的毯子也是灰色的，扶手椅和沙发则是蓝绿色。主卧在某种程度上是对城市的赞颂，以城市公园的壁纸为特色。

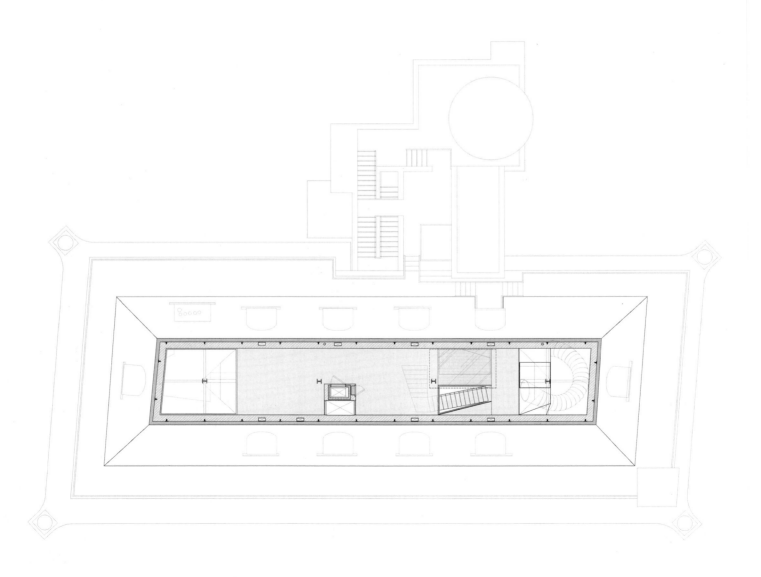

NEW YORK LIVING · NEW YORK CITY, NY

THE STARLIGHT

星光

设计公司	设计师	摄影师
DESIGN COMPANY	**DESIGNER**	**PHOTOGRAPHER**
Drake Design Associates	Jamie Drake	Marco Ricca

When designing the model apartment of One57, Jamie Drake was asked for something that would appeal to a broad array of prospective buyers, something gender neutral. They welcomed color but didn't want to be overwhelmed by it. The design is, naturally, all about the views. The foyer introduces you to the tone of the apartment with all of the round shapes, giving off almost a celestial experience of planets and meteor showers. The cosmic theme is carried throughout the apartment, in the luminous wall treatments, metallic details, lacquered woods, and shiny fabrics such as silk, velvet and satin.

在设计位于One 57大厦的这间样板公寓的时候，杰米·德雷克受命做一些能够广泛吸引潜在客户的中性化的东西。设计团队热衷于色彩却又不想受制于它。这里的设计都自然而然地与观景相融合。玄关将人们带入公寓，公寓内所有圆形图案几乎发散出行星与流星雨般的光芒。夜光的墙面处理、金属细节、漆涂木材以及丝绸、绒布、缎面等光面布料的运用将宇宙主题展现出来。

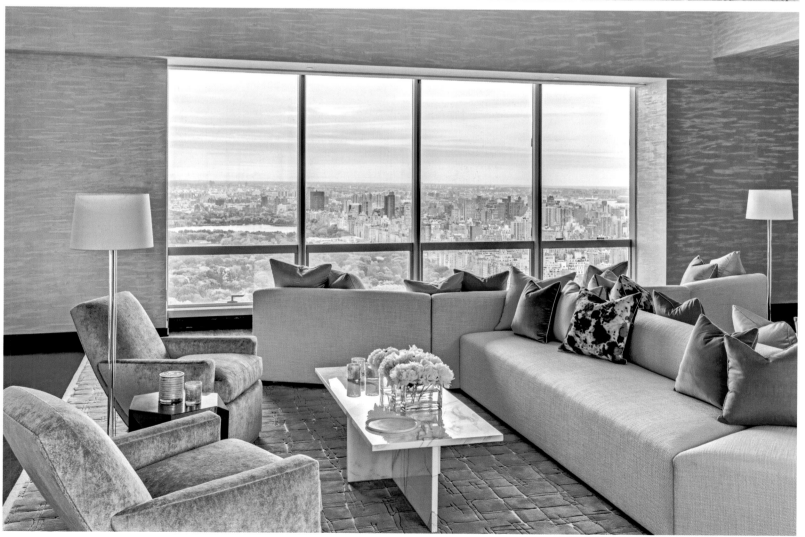

NEW YORK LIVING · PORT WASHINGTON, NY

COLORFUL FAMILY RESIDENCE

缤纷家居

设计公司
DESIGN COMPANY
Baltimore Design Group

设计师
DESIGNER
Keith Baltimore

摄影师
PHOTOGRAPHERS
Ric Marder Imagery and Pablo Corradi

Interior designer Keith Baltimore took what was a typical colonial home and transformed it, inside and out, into a wonderfully colorful and lively home for a busy and energetic family. Even the pets are comfy here (several dogs and fish)! but glamour still resonates from Keith's designs.

Starting with the entry foyer, Keith chose a striking hand-painted harlequin pattern containing the colors he used throughout the house, since good interior design seeks to integrate a broad color palette. Separating the rooms are classic molding and architectural features to define spaces. Stained wood floors throughout are elegant and easy to clean.

室内设计师基斯·巴尔地摩选择了一个典型的殖民主义风格住宅，并从内到外将它改造成了适合一个忙碌而充满活力的家庭居住的多彩又温馨的住宅。就连宠物（几只小狗和鱼）住在这里都很舒服，基斯的设计果然很有魅力。

从大厅入口开始，基斯就选择了一个引人注目的手绘模式，包括整个房子的取色，毕竟好的室内设计讲究颜色的搭配。传统造型和建筑特色决定了不同空间的布局。地面的着色板很高雅，而且容易清洗。

137

A basement containing the gym and play room has strong contrast colors to brighten and energize the space with bold orange and sky periwinkle blue. Fun abounds with a ping-pong table, pin ball machine, and stacks of games discreetly tucked away. The den has a huge television, electric fireplace, fish tank, and super deep, large sofa.

地下室包括健身房和游戏室,橙色和天蓝色的鲜明对比使空间更明亮更富有活力。打乒乓球,玩弹珠,都是很有趣的游戏。书房有大电视、电壁炉、鱼缸和很大很深的沙发。

145

The teen-age daughter's bedroom and study area is crisp white with bright red accents on the walls and red leather drawer pulls. As she grows, this room will still feel sophisticated with its Zen-like ambiance. A warm music room, with a bit of an "old world" vibe, is where the family clusters for a sing-along tapped out on the antiqued piano, with the dogs curled up at their feet.

年轻女儿的卧室和书房是白色的，但墙上有些醒目的红色装饰品，抽屉上也有红色的皮革把手。即使随着女儿的成长，这个房间仍然会很精致并富有禅意。温暖的音乐室，有点旧时的感觉，一家人随着钢琴的节拍唱着歌，小狗蜷缩在他们脚下，多么温馨的画面。

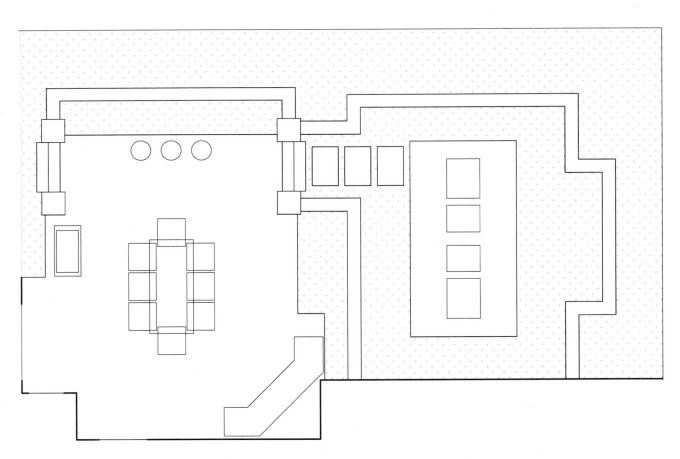

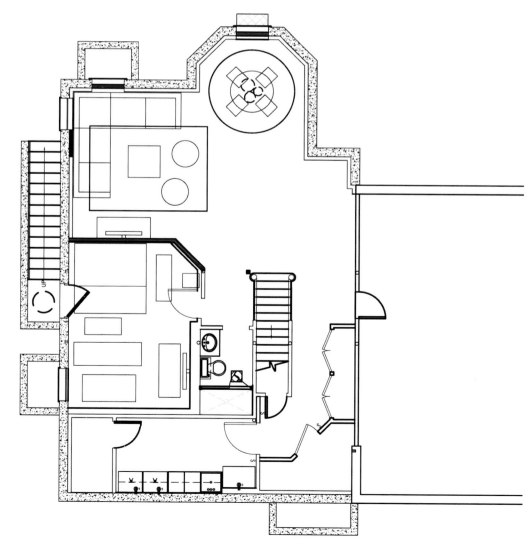

153

NEW YORK LIVING · SCARSDALE, NY

THE ELEGANT RUSTIC HOME

乡村雅居

设计公司
DESIGN COMPANY
bSTUDIO Architectural Design, LLC

设计师
DESIGNERES
Breanna Carlson & Brynn MacDonald

摄影师
PHOTOGRAPHERS
Breanna Carlson & bSTUDIO Team

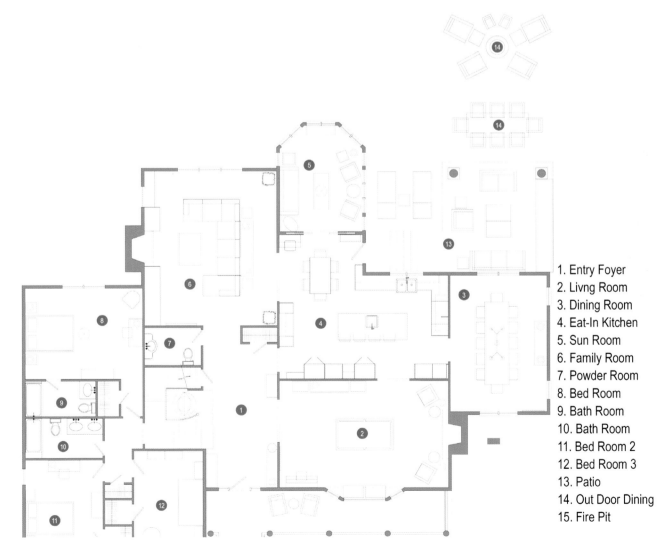

1. Entry Foyer
2. Livng Room
3. Dining Room
4. Eat-In Kitchen
5. Sun Room
6. Family Room
7. Powder Room
8. Bed Room
9. Bath Room
10. Bath Room
11. Bed Room 2
12. Bed Room 3
13. Patio
14. Out Door Dining
15. Fire Pit

This 6-bedroom, 5 bathroom home is designed with all of the modern conveniences. A young family from Manhattan had come to expect living in the city but with the added benefit of substantially more space and a yard that drew them the Scarsdale. Traditional elements throughout the home were incorporated into a French country design aesthetic for a casual transitional look. Children' bedrooms, bathrooms, and study are on the mezzanine level and above them are guest quarters and the master suite with his and hers walk-in dressing rooms, a private office, and en-suite master bathroom. Each room has a unique combination of materials and furniture providing differing textures and colors appropriate to the program of the space. The lower level of the home is more utilitarian with a mudroom, gym, boxing area, children's playroom, wine cellar, and direct access to the garage.

这个房子有六个卧室，五个浴室，并配有所有的现代便利设施。来自曼哈顿的一个家庭想搬来斯卡斯代尔居住，因为这个房子有更大的空间和一个庭院。贯穿整个空间的传统元素纳入了法式设计美学，使得整个过渡看起来比较随意。孩子的卧室、浴室和书房在中层，上层是客房、带有更衣室的主卧、私人办公区和主卧的浴室。每个房间都有不同的物品和家具搭配，不同的纹理和颜色也很适合各个空间的结构。下层更实用，有寄存室、健身房、拳击区、儿童游戏室和酒窖，也有入口直通车库。

The main entrance opens onto an inviting foyer which sets the tone for the home with a massive rustic industrial chandelier over the staircase, an antique cart console, damask hand-printed grasscloth wall covering, and a custom damask stair runner adding pattern and warm soft tones. This area connects the living and entertaining spaces with the more private sleeping areas on the floors above and utility rooms below. The front living room was once a formal family room which has been converted into a game room complete with a pool table, bar, piano, oversized seating, and fireplace.

　　大门入口直通门厅，门厅奠定了整个房子的基调，楼梯上方有一个巨大的乡村工业风的吊灯，通道里有一个古董车控制台，墙上有手工锦缎印花墙纸，阶梯上铺着定制的锦缎，这些都营造了一种温暖柔和的格调。这段楼梯连接着休息区和娱乐区，楼梯以上是私人休息区域，楼梯以下是比较实用的房间。游戏室是由一个普通的家庭活动房改造成的，里面有台球桌、酒吧、钢琴、超大号的座椅和壁炉。

A formal dining room is directly connected to the kitchen with a farm table for twelve and stunning wood chandelier. Hand-printed fabrics were sewn into window treatments for the living room, dining room, and kitchen with coordinating patterns and colors. The sunroom extends into the backyard through a set of French doors adjacent to the kitchen and features a loveseat, lounge chairs, and window treatments in a warm blue palette.

餐厅与厨房是连通的,里面有一张能容纳十二个人的木桌子和一个引人注目的木头吊灯。起居室、餐厅和厨房的窗帘上都缝有手工印花纺织品。日光室通过厨房旁的法式大门延伸到后院,里面有双人椅、躺椅、蓝色的窗帘。

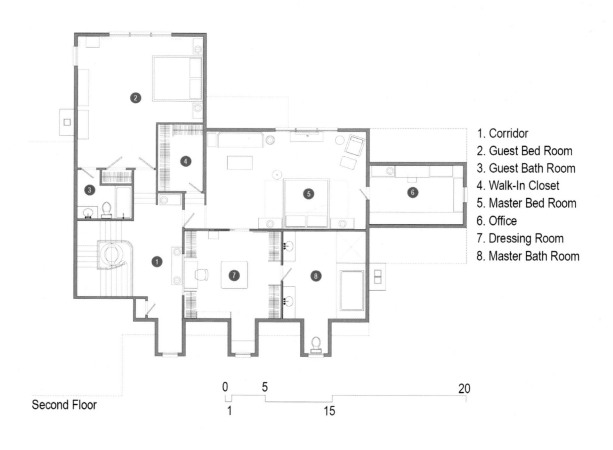

1. Corridor
2. Guest Bed Room
3. Guest Bath Room
4. Walk-In Closet
5. Master Bed Room
6. Office
7. Dressing Room
8. Master Bath Room

Second Floor

165

NEW YORK LIVING · SOUTHAMPTON, NY

DELICATE COTTAGE

精致小屋

设计公司	设计师	摄影师
DESIGN COMPANY	**DESIGNERS**	**PHOTOGRAPHER**
Design House	Suzanne Caldwell, Maria Greenlaw	Costas Picadas / Dluximage

Suzanne Caldwell is the co-owner of Design House with Maria Greenlaw, they renovated her house in Southampton village.

苏珊娜·考德威尔与玛丽亚·格林劳是Design House公司的共有人，她们在南安普敦的村庄翻修了苏珊娜的家。

The main house is about 186 square meters, the barn is about 111 square meters and the shed is 46 square meters. La shed's the latest addition that designers love, that was an old decrepit garden shed there. That they decided to replace, and of course one thing turned into another. Designers have turned it into a little outdoor summer house where guests stay in the summer, or they have dinner parties there, it's just a great versatile space.

Le Shed is outfitted with custom built in day beds, which serve for entertaining, seating purposes, and also can be made up for summer guests. As for the pillows, the leaf pattern is a pretty classic Peter Dunham textile. The other pillows are authentic handmade Mexican embroidered tiles.

房子主体面积186平方米，谷仓约111平方米，小屋为46平方米。设计师非常喜欢新增设的小屋，它过去是一间老旧的花园棚屋。她们打算替换掉它，将它变成一间户外避暑小舍，夏天的时候客人们可以在里面驻足停留或举办晚宴，这是一个多功能空间。

小屋配备了定制的沙发床，可供休闲就座之用，也可以供夏季的客人们使用。叶子图案的抱枕采用经典Peter Dunham面料，其他几只抱枕是墨西哥纯手工刺绣。

As for the main living room, on the English oak butler server, there is a collection of natural curiosities. The skull actually comes from Suzanne's brother-in-law's farm on the North Fork. The painted sticks and shells that designers have collected is a natural curiosity.

主客厅的英式橡木柜像是一处自然珍品收藏。头骨摆件来自苏珊妮姐夫的北福克农场。漆画的树枝、贝壳是设计师的藏品，这些都是自然珍品。

The sun room was originally a little sleeping porch. Designers turned it into a really intimate space for reading or watching TV and having coffee.

阳光房原先是一个小凉台。设计师将它转变成可供阅读、看电视或喝咖啡的私密空间。

The garden is pretty much surrounded by box wood, with the idea of winter gardening. You still have a lot of green ivy and box wood and structure in the winter. Then inside the box wood are more wild flowers and some are perennials. Designers are fortunate enough to be surrounded by ancient maple trees.

花园非常漂亮,四处环绕着黄杨,其构想是打造冬景花园,这样即使在冬天,也能看到大片绿色的常春藤和黄杨。黄杨之内有很多野花,其中有一些是多年生植物,非常耐寒。这里十分幸运地被古枫树环绕。

The dining room is outfitted with custom bookcases which house a rather large collection of books, covering subjects from design books, to history, to cooking, to nautical terms, to classic literature and travel books.

餐厅配备定制书橱，摆放着大量藏书，内容涵盖设计、历史、烹饪、航海、古典文学、旅游等等。

The alcove in the kitchen is created by the staircase that goes up overhead out of the dining room. The pantry was where the original foyer was. They've updated it and brought the laundry, washer and dryer upstairs, and outfitted it with custom shelves. The wallpaper is an old Pier Fre wallpaper that found in France.

厨房里的壁龛是由餐厅外向上延伸的楼梯建成。原来的玄关被用作餐具室。设计师们将其更新，在楼上放入洗衣机、洗涤器和烘干机，配备了定制的架子。古老的Pier Fre壁纸是从法国寻来的。

The master bedroom features textiles on the curtains and headboard and bed skirt on the vintage antique slipper chair. The Chinese double happiness lamp sits on the knoll tulip table, which adds a freshness to the room. The custom lampshades were actually made out of an additional bedspread that matches the bedspread that's on the bed. Lined in pink silk.

窗帘、床头板、复古矮脚椅的椅裙上的织物是主卧室的特色。成对的中式喜灯放置在郁金香小桌上,为房间增添了新鲜感。定制灯罩实际上是由与床单匹配的额外面料制成,采用粉色真丝线脚。

The bathroom wallpaper features an Ottoman motif designed by David Hicks in the 70s.

浴室的壁纸是来自70年代David Hicks的设计,以土耳其人物造型为特色。

NEW YORK LIVING · MANHATTAN, NY

HOME AND NATURE
自然与家

设计公司
DESIGN COMPANY
Andrew Franz Architect PLLC

设计师
DESIGNERS
Andrew Franz and Jaime Donate

摄影师
PHOTOGRAPHER
Albert Vecerka/Esto

In Manhattan's landmarked Tribeca North area, the 279 square meters top floor and roof of an 1884 caviar warehouse are reconceived as a warm and welcoming residence with large open entertaining zones and a fluid connection with the outdoor environment.

在曼哈顿地标性的特里贝克北区，这个1884年的顶层面积达279平方米的鱼子酱仓库经过重新构思设计，成为了一座温馨且受欢迎的住宅，拥有大型开放式的娱乐区，直通户外。

OUTDOOR CONNECTION

The residence is transformed by a relocated mezzanine where a sunken interior court with a retractable glass roof connects to the planted green roof garden above. This gesture of subtracting volume from the interior brings the outdoors into the primary living zones. The roof showers the spaces with natural light. When open, ample air flow enters what was once a poorly ventilated and dark loft. By night, the court acts as an internal lantern illuminating the loft below.

户外连接

该住所由一栋重塑的夹楼改造而来，下沉式室内球场、折叠式玻璃屋顶与上方绿植屋顶花园互相连通。室内的这种节省空间体积的设计将户外环境带入了主要的生活区域。屋顶射入的自然光沐浴着室内空间，充足的气流迎面扑来，谁能想到这曾是空气流通不畅、光线昏暗的公寓。在夜里，球场起到室内照明的作用，照亮了下面的公寓。

HISTORIC DIALOGUE

Embracing the building's industrial past, a visual discourse between new and old is devised through insertions of modern materials along with restored or reclaimed materials from the loft. A custom steel stair repurposes timbers from the old roof joists as treads and landings.

历史对话

　　基于这栋建筑工业化的过往，通过现代材料的修复、再生材料的融合，实现了新旧之间的生动转变。定制的钢梯重新定义了用作踏板和梯台的木梁的用途。

SUSTAINABILITY & HEALTH

To add to its sustainable nature, new and energy-efficient mechanical systems and appliances are employed. The project reclaims and reuses loft materials while bringing in new, locally sourced products including the appliances, retractable glass roof, architectural metal work, and cabinetry. The new roof terrace utilizes reclaimed bluestone pavers and a majority of native plant species that require little water while insulating the environment below.

可持续性&健康

为增加其可持续性，设计师采用了新型节能力学系统和家用电器。该项目回收再利用公寓的原始材料，同时引进当地新型物料，比如家用电器、折叠式玻璃屋顶、建筑金工及细工家具。新的屋顶平台铺有再生青石和大量原生植物，即使与下方环境隔离，也几乎不需要水。

NEW YORK LIVING · SHELTER ISLAND, NY

WALKING ALONG THE BEACH

漫步沙滩

设计公司
DESIGN COMPANY
Sara Story Design

设计师
DESIGNER
Sara Story

摄影师
PHOTOGRAPHER
Jonny Valiant

For inspiration, Sara Story did not need to look far. Her goal was to ease the house into the topography, and make it feel private, while maximizing the scenery.

Most rooms, including the upstairs bedrooms, have ocean views. The main level, which encompasses the kitchen, dining and living rooms, is also connected to the exterior by a 78-foot terrace, overlooking the sea and creating an indoor-outdoor circulation loop. Story wanted the interiors to be very neutral and for the focus to be on the spectacular scenery outside. She used the sky, beach, water and trees as her palette. The flooring is white oak, a compliment to the casual chic beach décor. The walls are mostly a creamy hue with minimal moldings, serving as a backdrop to clean-lined transitional furnishings.

莎拉·斯托里并不缺乏灵感。她的目标是改变房子布局，使它更具私密性，并最大限度扩大观景。

很多房间都可以观海景，包括楼上的卧室。主楼层包含厨房、餐厅和客厅，通过一个78英尺的阳台与外部相连，俯瞰海面，打造室内外循环回路。斯托里希望室内色彩中性，使焦点集中在室外的壮阔美景之上，所以她以天空、沙滩、海水、树丛为调色板。地板运用白色橡木，与休闲时尚的沙滩装饰形成互补。墙壁大多采用奶油色调，线脚简约，作为利落的过渡家居背景。

In the serene sand-and-shell-hued living room, silk area rugs add warmth to the floors, linen curtain panels subtly frame expansive window-walls and delicate sheers filter sunlight as needed. The living room is a large rectangular space, so Story broke it up by creating two separate sitting areas. The family uses the more casual area centered around the fireplace.

客厅运用沉稳的的沙滩贝壳色调，丝质装饰地毯为地板增添了温馨感，亚麻窗帘巧妙地框定广阔的窗与墙，精致的薄纱在需要的时候可以透过阳光。客厅是一个大型矩形空间，斯托里为拆散布局，打造了两间独立起居室。家属使用环绕壁炉的更为休闲的区域。

There is a surge of saturated color by way of the lime green chair cushions and curtain panels in the dining room, as well as the peacock blue wallpaper hand painted with a botanical theme. It works perfectly for more formal dinner parties at the beach. The resulting house is harmonious in design and at one with its environment.

设计师大量运用饱和色，通过餐厅内的石灰绿椅垫和窗帘以及植物主题手绘的孔雀蓝壁纸显现出来。它方便沙滩上更正式的聚餐。最终达到设计与环境和谐统一的目的。

Seating is upholstered in comfortable fabrics with stain-resistant finishes. Patterns are used sparingly, either simple geometric variations or abstract organic shapes.

落座区用舒适布料与抗污表层重新装饰软垫。运用了少量图案花纹，有简单的几何图样，也有抽象的有机图形。

195

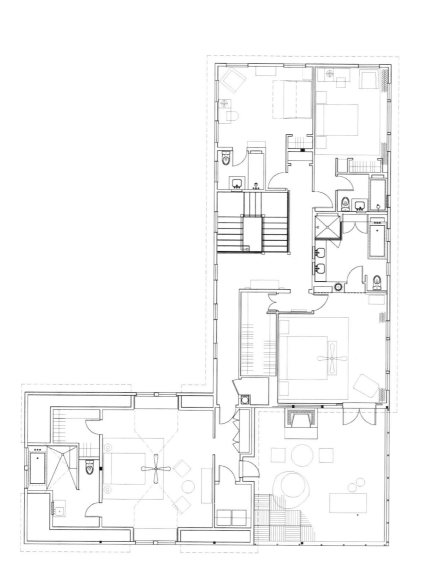
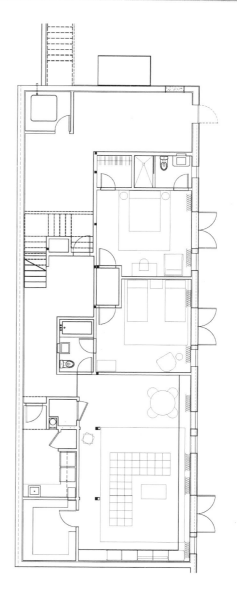

NEW YORK LIVING · NEW YORK

THE DANCING FAIRY
舞动的精灵

设计公司
DESIGN COMPANY
Fawn Galli Interiors

设计师
DESIGNERS
Fawn Galli

摄影师
PHOTOGRAPHER
Costas Picadas/ Dluximages

This is a loft duplex apartment. It is designed for a young couple with 3 young children, it has 4 bedrooms, 4 bathrooms, 8 rooms and 1 balcony. The challenge is to make an open space more traditional and family friendly. The clients wanted a comfortable glamorous space. The vision for the apartment is a mix of Art Deco in 50s. The color scheme for the main living dining is various shades of blue and green.

这是一套复式公寓。屋主是一对年轻夫妇，有三个小孩，公寓内共4间卧室、4间卫生间、8个房间及1个阳台。设计挑战是将这个开敞空间变得更加传统、居家。客户想要一个舒适且迷人的空间。这里的愿景是装饰艺术与50年代风格的混搭。主生活区、用餐区的配色选用蓝色与绿色的各式渐变。

Many of the bathrooms are dressed up with bold wallpapers. Designers used different elegant woods and plexi to create a glamorous but funky interior.

卫生间采用了各种风格大胆的壁纸作为装饰。设计师用不同讲究的木材及树脂玻璃打造了迷人时髦的室内空间。

NEW YORK LIVING · HAMPTONS, NY

BACK TO NATURE
回归自然

设计公司	设计师	摄影师
DESIGN COMPANY	**DESIGNER**	**PHOTOGRAPHER**
Fawn Galli Interiors	Fawn Galli	Costas Picadas / Dluximages

Houses are like people: sometimes their exteriors reflect their interiors, and sometimes they don't. And when they don't, that's when things get interesting. Take, for example, this handsome farmhouse, somber and straight-laced on the outside, lighthearted, casual, and whimsical on the inside, for which the nod goes to Fawn Galli, New York-based interiors specialist.

房屋就像人一样：有时候，室外反映着室内，有时却不然，有趣的事情便会发生。例如，这间大方的农舍，外部虽然很普通、一板一眼，内部却轻松随意、异想天开，其设计理念出自纽约当地室内设计专家弗恩·加利。

207

The clients asked Galli, who had created the family's Manhattan home, to accompany them on their vacation home-hunting trek. Fawn Galli always looks for surprises, twists on what's traditionally expected. With this objective in mind she had ready and active conspirators in the owners of the house. Parents with a blended family of children aged 2-19, their only submission to tradition was to the practicalities of comfort and childproofing. To accommodate both these goals the house is devoid of anything antique or precious. Fabrics are sturdy cottons and linens; the color palette of cool blues is light and playful; and the furniture leans towards the indestructible.

加利曾为这个家庭打造了曼哈顿的住所，客户邀请他同去狩猎度假屋。弗恩·加利一向热衷有别于传统的惊奇惊喜，正因为如此，她决意加入到这个项目的设计中。这是个重组家庭，孩子在两岁到十九岁之间，他们仅在舒适性及防止儿童损坏方面向传统做出了妥协。为实现这些目标，房内没有放置任何古董或珍贵之物。选用结实的棉麻织物，清爽的蓝色系轻盈而愉悦，家具坚固牢靠。

Indeed, a gigantic resin boulder anchors the twin living room sofas and the breakfast table has the classic Eero Saarinen steel base, which is resistant to any amount of toddler wear and tear. The table is circled not by the expected Saarinen-designed-to-match Tulip chairs, but by 1950s-ish wood flea market finds, painted the owners' favorite turquoise color, which look like they would be at home in a playroom. In another bit of playfulness, the stair treads are covered in different pieces of salvaged Indian cotton carpeting.

Only in the serious, gloss purple dining room-designed, to create a sense of "occasion"-does the house take a left turn from its overall lightness. But then the handcrafted, offbeat, brass-and-topaz chandelier brings it back home.

　　连固定在客厅两张沙发与早餐桌的巨大树脂卵石都配有足以抵抗孩童的任意磨损的经典的沙里宁钢制底座。桌子并没有配有沙里宁配套郁金香座椅，而是选用从20世纪50年代的木材跳蚤市场淘来的桌椅，涂上屋主最喜欢的蓝绿色，让他们仿佛置身于一个娱乐室般的家。楼梯踏板上的印度棉线地毯也是此案的另一处趣味所在。

　　只有在庄重的光漆紫色餐厅，才能打造出一种有别于整体亮度的"场合"。手工制作的另类黄铜黄宝石吊灯将光芒重新带回家中。

NEW YORK LIVING · BROOKLYN, NY

CALM AND ELEGANT
静谧与优雅

设计公司	设计师	摄影师
DESIGN COMPANY	**DESIGNER**	**PHOTOGRAPHER**
Willey Design LLC	John Willey	Robert Granoff

When a New York City townhouse needed a transformation, designers were proud to retain the 100-year-old charm while giving it a youthful lift. They created a combined dining and living room with a corner seating area to provide multiple uses for today's family. The walls are combed and tinted plaster and set against a sky-blue ceiling. Cream velvet upholstery, geometric patterns, and a delicious grey and blue rug complement a striking cluster globe chandelier, modern art, and unexpected vintage elements. The quietly luxurious room evokes a calm and sensible elegance.

当得知这座纽约别墅需要进行改造的时候，设计师感到很荣幸，因为这样可以既保留其百余年来的历史魅力，又能为它增加年轻朝气。设计团队通过一个角落座位区将客厅与餐厅合二为一，为如今的家庭提供多功能用途。墙壁由彩色石膏打造，倚靠着天蓝色的天花板。乳白色天鹅绒软装、几何图案、怡人的灰蓝地毯与醒目的球形吊灯、现代艺术及惊艳的复古元素形成互补。低调而又奢华的房间展现出一种静谧优雅之美。

235

NEW YORK LIVING · NEW YORK CITY, NY

A LITERARY JOURNAL
艺术之旅

设计公司
DESIGN COMPANY
bSTUDIO Architectural Design, LLC

设计师
DESIGNERS
Breanna Carlson, Brynn MacDonald and Tilton Fenwick

摄影师
PHOTOGRAPHER
Shaked Uzrad

1. Entry Foyer
2. Gallery
3. Office
4. Powder Room
5. Master Bedroom
6. Master Walk-In Closet
7. Master Bath
8. Living/Dining Room
9. Kitchen
10. Laundry
11. Family Room
12. Guest Bath
13. Child's Bath
14. Guest Bed
15. Child's Room
16. Yoga Room/ Gym
17. Storage

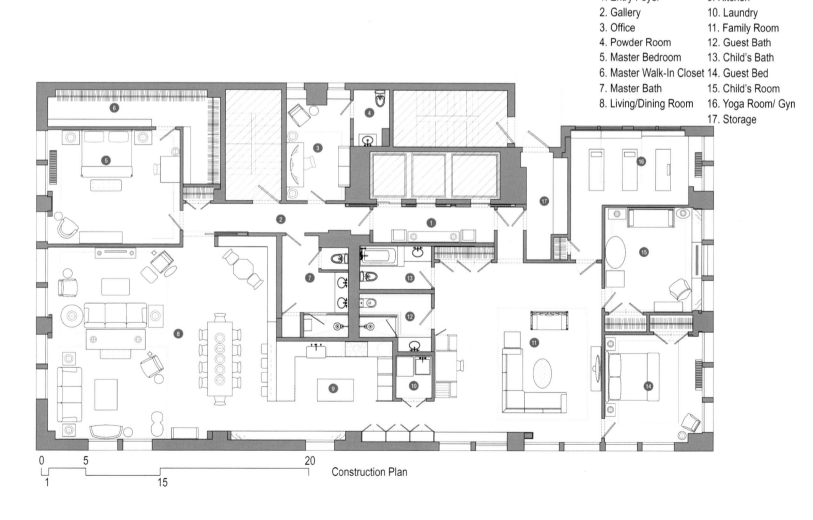

Construction Plan

With a spacious but anonymous floor plan as the foundation, this loft renovation was tailored to the couple's desire to have a space that effortlessly balances work and family. The demolition and construction plans were designed to create purpose-driven divisions of space with a comfortable and welcoming flow.

A home office with en suite powder room is located apart from the living, dining and family areas creating a self-contained space for focused work within the home.

Walls enclosing the kitchen were removed reconnecting the living, dining and kitchen areas and eliminating visual barriers along the corridor to the family room.

这个房子是以一张毫无特征的宽敞的平面图为基础为屋主夫妇量身定制的，以满足他们拥有一个能轻松平衡工作与家庭生活的空间的愿望。改拆和施工计划旨在目标驱动的空间分布和舒适受欢迎的氛围。

配有盥洗室的家庭办公室离客厅、餐厅及家庭区域很远，在家里打造了一个设备齐全、可以专注工作的空间。

厨房的围合式墙壁被拆掉，重新与客厅、餐厅及厨房连接，消除了从走廊到房间沿线上的视觉阻碍。

With a priority to open and reconnect the communal spaces, a closet was eliminated in the renovation. In its place, walls were redesigned as bookcases and cabinets and built-in window seats in the child's room double as toy storage.

A second office space is located in the family room allowing the couple a space to casually work while minding their young daughter. A reading bunk built in above the desk fosters a playful, kid-size quiet space for their daughter.

Indian-inspired patterns, tiles, and a vibrant color palette used throughout the loft reflect the family's Indian heritage and push the renovation beyond structural changes to fully embody the life and personality of the loft's inhabitants.

为了优先展开并重新连接公共空间，重建时拆除了一个壁橱。取而代之的是经过重新设计的书架、陈列柜以及儿童房的嵌入式窗座，还可用作玩具储藏室。

另一个办公区位于家庭室，因此这对夫妇可以边工作边照看小女儿。上床下桌式的设计为他们的女儿提供了可以玩耍的安静空间。

受印度风格灵感启发而来的图案、瓷砖和鲜艳的色彩搭配，反映出这个家庭的印度传统，使这次翻修凌驾于结构变化之上，充分体现了居者的生活与个性。

NEW YORK LIVING · UPPER EAST SIDE, NY

RETRO AND MINIMALIST

复古与简约

DESIGN COMPANY
Design House

DESIGNERS
Suzanne Caldwell, Maria Greenlaw

PHOTOGRAPHER
Costas Picadas / Dluximages

This project is located on the Upper East Side, designed by Maria Greenlaw and Suzanne Caldwell from Design House. A young urban couple, and one child and one cat live there.

本案位于纽约上东区，由Design House公司的苏珊娜·考德威尔及玛利亚·格林劳设计的。一对年轻的城市夫妻带着一个小孩和一只猫生活在这里。

In the front entry, there is a horse hair wall covered by Phillip Jeffries, the light fixtures is made by Porter Romano. The white console in the front entry is a Tommi Parzinger circa 1950s piece and the bench opposite it, the black bench in the little alcove is by Harvey Probber with a lithograph hanging above it. Above the console is a French style mirror circa 1940. In the living room area they have the drapery fabric which is a custom embroidered fabric by Holland and Sherry. A custom acrylic cocktail table and a pair of leather and rod iron benches by Paul McCobb. The custom wall unit is designed by Design House and made by Manhattan Cabinetry with horn hard ware by Ochre and a hand woven wool rug by Beauvais carpet.

前面的入口处采用了Phillip Jeffries的马毛壁布以及Porter Romano的照明设备。白色漆面的接待台出自20世纪50年代的Tommi Parzinger，对面小壁龛内的黑色长凳是Harvey Probber所做，上面有一幅石版画。接待台上方是20世纪40年代的法式镜子。客厅的装饰织物是Holland和Sherry定制的绣花面料。亚克力鸡尾酒台、一对皮革搭圆铁长椅由Paul McCobb定制。组合柜由Design House公司专门设计，Manhattan Cabinetry公司以Ochre角形硬陶及Beauvais carpet手工编织羊毛毯制成。

The chest table is a black lacquer custom piece. The dining table is also a custom table with a rod, iron base and a walnut top. The chairs are by David Iatesta. The pair of large light fixtures is Italian circa 1960. The mirrored console is a vintage piece as well in the dinning room. The alabaster lamps on top of the console are circa 1960s with custom feather lamp shades was made in the UK.

矮桌是黑色漆面定制。餐桌则是由踏杆、铁质基底及胡桃木桌面定制而成。餐椅是David Iatesta设计的。那对大型照明灯具为20世纪60年代的意大利风格。餐厅里的镜面操作台同样也是复古的老物件。操作台上摆放的雪花石膏灯具来自20世纪60年代，羽毛灯罩是在英国定制的。

Glass light fixtures over the bar are hand blown by an artist named Caleb Simon and the wall sconces near the game table are by Vaughn Lighting. Then in the library they have a custom sectional by Avery Boardman in New York and with fabric by Holly Hunt. Cocktail table is also made by Holly Hunt. Cushion fabric on the window seat is a Hermès fabric. Again, custom cabinetry is designed and made by Manhattan Cabinetry.

吧台之上的玻璃灯具由艺术家Caleb Simon手工打造，游戏桌旁边的壁灯由Vaughn Lighting打造。书房中有从纽约定制的Avery Boardman组合柜及Holly Hunt的织物。鸡尾酒台也是由Holly Hunt打造。窗座的软垫采用爱马仕面料，柜子由Manhattan Cabinetry设计制造。

Then, in the master bedroom designers have a custom head board that they made for the client. Night tables are by Todd Hase. The rock crystal lamp in the bedroom with a custom silk shade is also from Vaughn Lighting. The chest is a piece that was purchased for a different apartment and moved into this apartment with the client, it has ebony and ivory inlay. The drapery fabric is a silk and, here it is, the Art Deco chest of drawers has walnut, fruit wood, and ebony inlay. It's French circa 1930.

These amethyst geode lamps are designed and made by Design House.

设计师在主卧室为客户定制床头板，采用Todd Hase床头柜。摇摆的水晶灯与定制丝绸灯罩由Vaughn Lighting打造。矮桌是在另外一间公寓购买、随着本案客户一起搬入这间公寓的，它上面镶有黑檀和象牙。丝绸作为装饰织物，五斗橱来自20世纪30年代的法国，带有胡桃木、果木及象牙镶嵌。

紫水晶灯具由Design House公司设计并制造的。

NEW YORK LIVING · UPPER EAST SIDE, NY

BRANDISH YOUTH
舞动的青春

设计公司
DESIGN COMPANY
Willey Design LLC

设计师
DESIGNER
John Willey

摄影师
PHOTOGRAPHER
Robert Granoff

This apartment is renovated by John Willey, who is accustomed to the unpredictable events that unfold in the world of white-hot New York real estate. The apartment was spacious but stodgy. It was quite heavy, with lots of chocolate browns and tans. It had big swag draperies that closed off the best views, such as the Metropolitan Museum of Art across the street. The goal was to lighten it up and give it a more playful and youthful character. They love color and a fun, jazzy quality. The couple wanted to use the furnishings from their previous apartment, a mix of custom-designed pieces and a treasure of vintage furniture from Karl Springer, André Arbus and Edward Wormley. But given that they were starting over, they decided to switch up the color scheme.

这间公寓由约翰·威利改造，他擅长于处理纽约炙手可热且不可预知的地产项目。这套公寓原本很宽敞但是枯燥乏味。很多巧克力棕褐色，过于沉重。大量垂饰物阻隔了极好的风景，比如街对面的大都会艺术博物馆。设计目标是点亮这个空间并使其更加赏心悦目、富有朝气。屋主喜欢有趣、花哨的颜色。他们希望利用之前公寓中的家居，包括定制家具的混搭以及卡尔·斯普林格、安德烈·阿尔比斯和爱德华·沃姆雷的古董家具。然而考虑到正重新开始，设计团队决定更换配色方案。

For the new apartment, Willey chose a palette of gray, black, pale blue, ivory, and pops of cranberry. He bleached all of the floors to a beautiful Scandinavian gray to get a dreamy, in-the-clouds effect that makes the apartment feel calm and serene.

With the help of builder Ted Thirlby and his team, Willey reworked the floor plan slightly to better suit the family's needs. The team also cleaned up the architecture to bring in the light from Fifth Avenue and to give the apartment a more youthful energy. The design throughout was intended to best showcase the couple's ever-growing art collection, which includes pieces by George Condo, Louise Bourgeois and Harland Miller.

　　威利为新公寓挑选了灰、黑、淡蓝、象牙白及蔓越莓的配色。设计师漂白所有地面，换成斯堪的纳维亚灰色，达到一种梦幻、如在云间的效果，令公寓看起来平静安详。

　　在建筑商特德·瑟尔比及其团队的帮助下，威利为更好地配合家庭需求，稍稍更改了平面图。团队对这里进行清理，使第五大道的阳光映入室内，令该公寓更加年轻有活力。这里的设计一直意图展示这对夫妻与日俱增的艺术藏品，包括乔治·康多、路易斯·布诺瓦及哈兰德·米勒的作品。

Perhaps the most labor-intensive project is taking the apartment's somewhat dreary entry hall and transforming it into a striking gallery featuring broad stripes of limestone in smoke and champagne, with matching faux paint that continues the striped effect up the walls and across the ceiling. The goal is to create sort of an abstracted Chinese puzzle box for a 'wow' effect when you walk into the apartment.

In the end, the entry gallery—and the entire home—take on a visually distinctive yet serene vibe. It's all cool colors. When you walk in, it's very calm and soothing, which is the entire point of an apartment in New York City. You want to have a dip in an oasis after coming in from stressors of the crazy city.

或许劳动最密集的项目是将该公寓略微沉闷的入口大厅改造成醒目的走廊，在烟气缭绕与香槟芬芳中，以石灰石的宽条纹为特色，搭配人造漆，在墙壁和天花之上继续使用条纹效果，意在打造一种抽象的中式魔方，令人们在进入公寓之时感到惊艳。

最后，入口门廊与整个家都在视觉上呈现出与众不同却依然平静安详的氛围。所有一切都是冷色调。当你走进公寓，会感到安静平和，这是纽约市内公寓的整体要素。在这个疯狂城市的压力下逃脱，你想要的一定是一处可以徜徉的绿洲。

NEW YORK LIVING · NEW YORK CITY, NY

GORGEOUS SPACE
华丽空间

设计师
DESIGNER
Jeff Pfeifle

摄影师
PHOTOGRAPHER
Costas Picadas/ Dluximages

Pendant lights, wall art, and carefully selected decorative pieces are part of a New England style echoed throughout the apartment. The floors around the entrance of the apartment are terrazzo, reflecting a "deco-ish feel" found elsewhere in and around the apartment building. Jeff Pfeifle designed the apartment for himself.

Entering the apartment, you're literally transformed to another time and place. This feel is achieved with design elements such as mercury glass with three triptyches. The entrance, is more than just a spot for welcoming guests to the apartment, it's also a functional space occasionally used as a dance floor and the location for a bar area during various gatherings. Next to the coat closet and powder room is a telephone room with a little telephone and a chair set up so guests can have private conversations.

吊灯、墙壁艺术以及精选的装饰物都是新英伦风的组成部分，这些都被运用到此公寓的设计中。公寓入口处的地面是水磨石打造，反映出一种在这栋楼其他地方可见的"装饰般的感觉"。杰夫·法埃福既是屋主又是设计师。

进入公寓，不知不觉你会转换到另一个时空。这种感觉的实现归功于众多设计元素的运用，例如三联画水银玻璃。入口处不仅仅是接待客人的地点，它偶尔被用作舞池及各种聚会中的吧台区。衣柜和化妆间旁边是通话间，有一部小电话和一把椅子，这样客人们就有了私人交流的地方。

The paneling in the New England style living room is believed to be from Paris. The furnishings and artwork are an eclectic mix of different styles somehow forming a cohesive look. The decor in the living room includes two ceramic urns from Italy and three crystal vases. The floral art above one of the sofas was found in a garage in Bucks County. The painting over the fireplace was discovered at a little gallery in Saint Tropez.

The former TV room/den has been transformed into a sun room/office. The little garden on the balcony is a wonderful place. What's described as sort of an English garden complementary to the interior style of the apartment has a lot of boxwood. A few new trees and shrubs are added with each growing season.

新英伦风的客厅镶板来自巴黎。家居及艺术品是各种风格的混搭，无意中形成了一种聚合的外观。客厅中的装饰包含两只意大利的陶瓷缸和水晶花瓶，有一张沙发上方的花艺是从巴克斯镇一个车库得来，壁炉上的油画来自圣托佩斯一家小画廊。

从前的电视房/私室被改造成阳光房/办公室。阳台上的小花园美妙怡人。与公寓室内风格互补的英式园林运用了许多黄杨木，每个成长的季节都会增加一些新的灌木和乔木。

The nearly perfectly square New England style dining room adjacent to the kitchen is ideal place for enjoy cooking and entertaining. The room can easily fit four round tables to accommodate guests. The large painting, appropriately titled Big Leaves, was purchased in the 80s from a gallery in Palm Beach.

The narrow hallway leading to the bedrooms is made a little more special with bookcase wallpaper from Brunschwig & Fils. A few frame prints further create the illusion of a library to the casual observer. The two red lanterns are from English Country Antiques in Bridgehampton. As for the bedrooms, there's one master bedroom, one guest bedroom, and one maid's room.

近乎完美的正方形英伦餐厅毗邻厨房，那是享受厨艺的理想之所。这个房间可以轻松地摆开四张圆桌来款待客人。题为大叶子的大型油画购于80年代一家棕榈滩的画廊。

通往卧室的狭窄走廊有一点点特别，它摆放着Brunschwig & Fils的书柜壁纸。框架印花更进一步打造了书房的错觉。两盏红灯是源自汉普顿的英国古董。公寓内有一间主卧、一间客卧和一间佣人房。

NEW YORK LIVING · SOHO, NY

A CHARMING TEMPORARY HOME

魅力别居

设计公司	设计师	摄影师
DESIGN COMPANY	**DESIGNERS**	**PHOTOGRAPHER**
Robert Couturier Inc.	Joe Nahem and David Gorman	Tim Street Porter

279

During the weekend, when Couturier is in New York, he lives in a 233 square meters floor-through apartment, located right above his Soho firm's office. Although he moved there in 2000, he has utterly transformed the apartment on numerous occasions. He can try this space because he doesn't live there full time; his emotional life is not there, so he is less attached by the things that surround him. He thinks an apartment is a little bit as a suit, when you get tired of it you can easily change it to reflect your current taste.

However, the common thread through all those renovations is that Couturier's apartment has served as an urban crash pad that reflects his elective interest, including his passion for modern and contemporary design. In it's current incarnation, the apartment presents Couturier's open-concept living. Infinitely softer and more comfortable than an austere loft, it nevertheless blends different living functions. The sleeping area, work space, and bathroom are separated by a pair of low curved walls and all work together as one big room.

当库蒂里耶周末在纽约时，他都住在一个233平方米的全层公寓里，这个公寓离他索和区的公司很近。尽管他2000年才搬到这里，却已经对这个公寓进行了多次改造。他之所以尝试了多次，是因为他并没有一直住在这里。他的情感生活并不在这里，所以很少被周围的事物感动。他认为一个公寓有点像一套西装，当你厌倦某套西装时，可以对它做适当的改变以反映你当时的品味。

然而，贯穿每次改造过程的常见思路是这个公寓只是库蒂里耶的城市临时居所，反映了他有选择性的兴趣，包括他对现当代设计的热情。这个公寓目前的设计体现了库蒂里耶开放性的生活，融合了不同的功能，比没有任何装饰的阁楼更柔和舒适。休息区、工作区和浴室被两堵弯曲的低墙隔开了，但都在一个大房间里。

Here in his city home, he has varied wonderful things from different eras: 18th Century English paintings, 20th Century photographs by Couturier's friend David Seidner, 18th Century French Deco rugs and a Deco-era zebra-sofa by Jaques Adnet. Ancient curiosities are mixed with works by contemporary artists and designers such as Missoni Patchwork vases by Stephen Burks. There are also a few family heirlooms, including cabinets by Alfred Porteneneuve that were originally designed for his grandfather's office.

库蒂里耶在这个公寓里收藏了多种不同时代的物品：18世纪英国的绘画，20世纪来自好友大卫·桑德纳的照片，18世纪法国的装饰地毯和杰奎斯·阿德内特设计的有斑马纹的沙发。还有很多与当代艺术家和设计师的作品结合起来的古玩，比如出自史蒂芬·伯克斯之手的米索尼品牌的拼接花瓶。当然这里也有一些传家宝，包括原本阿尔弗雷德为库蒂里耶祖父的办公室设计的橱柜。

284

NEW YORK LIVING · UPPER EAST SIDE, NY

THE HERITAGE OF LOVE
爱的传承

设计公司	设计师	摄影师
DESIGN COMPANY	**DESIGNER**	**PHOTOGRAPHER**
Cathy Triant Buxton T.D.Triant Inc.	Cathy Triant Buxton	Costas Picadas / Dluximages

This house is located on the Upper East Side of New York, right down from the Metropolitan Museum. It is the ground floor of a townhouse. Four rooms and a garden.

It's a townhouse that designer's father originally developed in the early 70's. It was a complete wreck when he bought it. He developed it into two apartments that they've lived in ever since.

本案坐落于纽约上东区，沿大都会博物馆向下。这是一栋联排别墅的第一层，有四间屋子和一个花园。

这栋别墅最初是设计师的父亲在70年代早期发现的，他买下这里的时候完全是一片残骸。他将它改造成两套公寓，从此一直住在这里。

The big painting over the sofa in the living room is George Necraponte.

客厅沙发上方的大型油画来自George Necraponte。

The other artwork in that room is Xceron.

屋内其他艺术品均出自Xceron。

There's a back garden, sort of a TV/library, kitchen, entrance, dining, and a living room. Then a bedroom and two baths on the first floor.

Then on another floor, you can see living room, entrance, dining room, library, kitchen, and then three more bedrooms.

They have a sofa covered in a white linen. Eyecott pillows. Two chairs. Two little French chairs covered in linen and velvet that my upholsterer did. Two little benches covered with gray sheepskin, which designer got on First Dibs. The rug is a David Hicks design which is a reference to the past since designer's father used to work with David Hicks in the 70s.

这栋别墅有一个后花园、影音室/书房、厨房、入口、餐厅及客厅，一楼还有一间卧室和两个卫生间。

另外一层可以看到客厅、入口、餐厅、书房、厨房，以及三个卧室。

沙发是白色亚麻面料，配上Eyecott的抱枕，两把亚麻配丝绒的小法式椅由家具商打造。两把灰色羊皮长椅是设计师在First Dibs得来。地毯出自David Hicks，这与过去有所关联，因为设计师的父亲在70年代就常常与之合作。

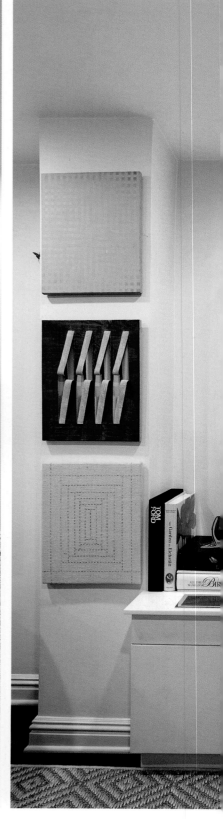

NEW YORK LIVING · SOUTHAMPTON, NY

A WARM AND COMFORTABLE RESIDENCE
温情家居

设计公司
DESIGN COMPANY
Fox-Nahem Associates

设计师
DESIGNERS
Joe Nahem and David Gorman

摄影师
PHOTOGRAPHER
Peter Murdock

303

The designers' mandate was to recreate an existing residence for an active family with three grown children, bringing it up to date through extensive renovations and additions, suitable for their lives today. They incorporated multiple public and private spaces for all members of the family, creating a warm and welcoming home, both elegant and comfortable that provides a suitable backdrop for their extensive and varied art collection. The goal was to make each room unique, but inviting, so that no part of the house would go unused.

设计师的任务是通过不断的修整将现有的住宅改造成一个适合有三个成年子女的家庭居住的住宅。他们合并了多个公共区域和私人区域以便全家人使用，创造了一个温暖又受欢迎，高雅又舒适的家，同时也可以展示多种多样的艺术收藏品。设计师的目的是使每个房间都独特且受欢迎，因此整个房子没有一处空间是闲置的。

Deftly using a variety of textures and materials, including pine plank paneling, blackened steel, lacquer finishes, Corian, teak, and cork, and mixing those with hand-woven textiles, leather, suede, and beautifully woven rugs, the designers created a serene environment equally suited to entertaining large groups as well as comfortable family living. The family room fireplace blackened steel mantle, lacquer shelves, and the blackened steel and lacquer hanging bar were custom FNA designs, as was the Corian and teak master bath vanities and surrounds. Much of the upholstery including the sofas for the living room, family room, and den, the master bedroom bed were FNA bespoke pieces specifically designed for this client. Palettes were kept relatively neutral, but incorporated judicious dosages of color handled in a confident and restrained way, so as not to be too decorative or compete with the client's art collection. The designers are fanatical about comfort and all of our upholstery, whether designed to look formal or casual, is always super comfortable.

通过巧妙地使用各种材质，包括松木板镶板、黑钢、漆制品、可丽耐、柚木、橡木，并把它们与手工纺织品、皮革、麂皮和漂亮的编织地毯结合使用，设计师们营造了一个既适合居住又能款待朋友的宁静的居住环境。客厅壁炉的黑钢壳、涂漆架、连接黑钢壳和涂漆架的吊杠，还有可丽耐的橡木洗漱台，都是福克斯·纳厄姆设计公司定制的。大部分家具，包括客厅、休息室和书房的沙发，主卧的床，都是福克斯·纳厄姆设计公司为客户定制的。调色相对中立，明智的颜色搭配虽然受了限制，但不失自信，这样一来，整个装饰没有掩盖客户艺术收藏品的风采。设计师们追求舒适感，因此他们的家具，不管是正式的还是随意的，都超级舒服。

311

NEW YORK LIVING · WESTBURY, NY

THE GLAMOUR OF ART
艺术之魅

设计公司
DESIGN COMPANY
Geoffrey Bradfield, Inc.

设计师
DESIGNER
Geoffrey Bradfield

摄影师
PHOTOGRAPHER
Sargent Photography

Today, in an era when so much has become global, some of the romance of hotels captured in film *The Grand Hotel* has drifted into oblivion. But with this apartment, Bradfield has revivified the spark of magic that materializes when international points of view coalesce to form a truly cosmopolitan tableau.

The art collected here—from Russia, Mexico, Spain, Great Britain and the United States—celebrates the spirit of convergence.

A reception room continues the living room palette, which is featured on an aubergine cut velvet custom sectional and another custom rug from Stark. The grisaille painting creates the illusion of a hall extension beyond. The flowers in a fluted Murano glass bowl from John Salibello offer yet another shade of lilac.

经典影片《大饭店》展现的浪漫风情在全球化的今天可能已经不复存在。但在这间公寓里，白爵飞重燃各种文化视点碰撞的神奇火花，打造出一幅真正的国际大都会风情画。

这里的艺术收藏来自俄国、墨西哥、西班牙、英国和美国，展现文化的包容精神。

接待室延续了客厅的配色，比如紫红色立绒呢定制的组合式家具，以及另一张Stark定制地毯。纯灰色画打造了大厅延伸的错觉。约翰·萨里拜罗Murano沟纹玻璃瓶中的鲜花给人另外一种紫色氛围。

The entry hall floor continues into the dining area, creating a subtly rich backdrop for a table laid with heirloom embroidered linens and Barry Flanagan's bronze Unicorn. The Ruhlmann Macassar ebony sideboard and French 1940s gilded bronze sconces add Deco glamour.

"Mantras are holy because you repeat them," says American artist Hunt Slonem, explaining one reason for the repetition of imagery in his paintings. Indeed, this two-dimensional mantra, depicting a fluttering lek of butterflies against a shimmering gold ground, brings an ethereal quality to the master bedroom.

入口大厅的地面延展至餐厅，为餐桌打造了精致丰富的背景，桌上铺着祖传刺绣亚麻，摆放着巴里·弗拉纳根的青铜独角兽。于耳曼·马佳萨的檀木餐具柜和20世纪40年代法国镀青铜烛台一起增添了装饰艺术魅力。

美国艺术家亨特·斯洛能说，"祷文之所以神圣，是因为人们一遍又一遍地重复"，这句话解释了为何他的画作中有重复的意象群。二维平面的曼陀罗倚靠着闪耀的金色地面，描绘了蝶儿翩翩的美景，为主卧室带来超凡脱俗的品质。

The cigar room's gray leather walls wrap the space in a neutral, yet masculine, way, allowing a bespoke Chesterfield sofa and club chairs in complementing velvets to take on a sculptural bearing. The gold Louise Nevelson work, from her Baroque phase, has a softer aspect than her better-known black sculptures.

雪茄室的灰色皮革墙面将空间包裹成中性偏阳刚风格,定制的Chesterfield沙发与丝绒座椅呈现出雕刻般的风采。露易丝·奈威尔森的巴洛克风格金色作品比她著名的黑色雕塑更加温和、易接近。

In the Empire-style library, an interesting conversation about essential beauty and the superfluousness of fashion. Bradfield designed the Brazilian heartwood reading table with white gold accents and collaborated with John Yarema to create the custom inlaid millwork.

帝国风的书房中进行着关于必要之美与非必要时尚的有趣对话。设计师白爵飞设计了人造白金表层的巴西心材书桌，而约翰·亚若玛则打造了内嵌木工。